THE LLANTHONY VALLEY

A BORDERLAND

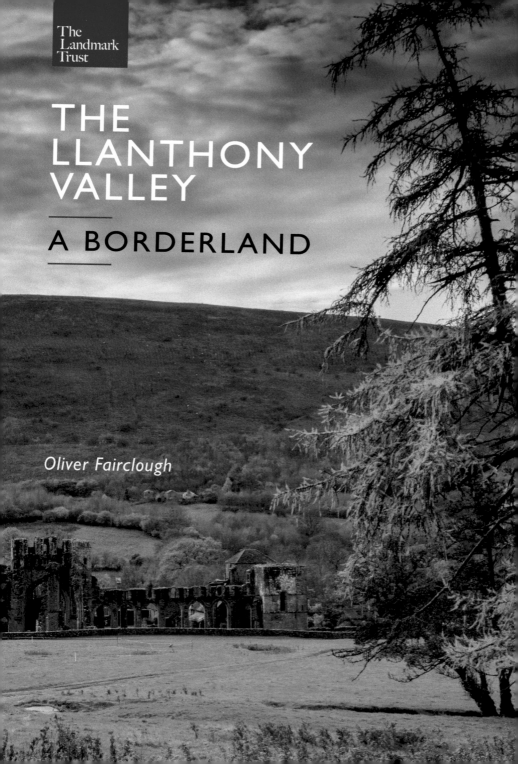

THE
LLANTHONY
VALLEY

A BORDERLAND

Oliver Fairclough

'These ruins derive a peculiar beauty from their situation
in the deep *Vale of Ewias*... Fertile in corn and pasture,
occasionally tufted with trees ... it is wholly encircled
by an amphitheatre of bleak and lofty mountains.'

– William Coxe, 1801.

This edition © The Landmark Trust, 2018
Text © Oliver Fairclough, 2018

First published in 2018 by
The Landmark Trust
Shottesbrooke
Maidenhead SL6 3SW
United Kingdom

www.landmarktrust.org.uk

ISBN 978-1-5272-2053-9

Edited by Caroline Stanford
Designed by Matthew Wilson
Printed by Lexon Group, Wales

10 9 8 7 6 5 4 3 2 1

The moral right of the author has been asserted.

The fonts used, both designed by Eric Gill, are Joanna MT and Gill Sans

Previous spread Llanthony Priory. © *Simon Powell*

The Landmark Trust

LLANTHONY
VALLEY &
DISTRICT
HISTORY
GROUP

**Funding raised by
The National Lottery**
and awarded by the Heritage Lottery Fund

**Cyllid a godwyd gan
Y Loteri Genedlaethol**
ac a ddyfarnwyd gan Gronfa Dreftadaeth y Loteri

Contents

Preface

In 2007, Cadw, the Welsh historic environment service, approached building preservation charity the Landmark Trust with a plea to save a perilously derelict fifteenth-century farmhouse called Llwyn Celyn that stood at the southern end of the Llanthony Valley. Fast forward a decade and a £4.2 million restoration project enabled by the Heritage Lottery Fund, and Llwyn Celyn and its site have been fully restored. The farmhouse is a holiday let available to all through the Landmark Trust, the threshing barn has become a flexible community and education space and bunkhouse, and an interpretation room has been created in the beast house. The project has enabled much else besides, and benefited from the support of many local residents and partner organisations.

In 2013, as it sought to learn more about this fascinating area, Landmark initiated a local history group. Much local knowledge already existed, and was cherished and shared, but there was then no focus or wider context in which to promote and explore it. By 2018, led by Pip Bevan and Douglas Wright, the Llanthony Valley & District History Group has grown to some 80 members, with well-attended monthly talks through the winter and guided walks in the summer. The Group is building an historical archive about the Llanthony Valley and its adjacent area. Some members actively research the social, economic and cultural life of the valley, its buildings and people; others contribute their own oral memories. This book is both a history and a snapshot of the valley in 2018.

Members of the Llanthony Valley & District History Group celebrate the news of a Heritage Lottery Fund grant towards the rescue of Llwyn Celyn in 2015.

Everyone in the Group has contributed to this book in some way, but we owe special thanks for its content to David Austin, Bridget Barnes, Adrian Betham, Pip Bevan, Edith Davies, Peter Davies, Vivien Davies, Diana Evans, John and Mary Evans, Caroline Fairclough (who also compiled the index), David Filsell, Jenny Francis, William Gibbs, Mary Griffiths, Miriam Griffiths, Chris Hodges, Russell James, Julia Johnson, Vicky Jones, David and Christine Knight, Sue Mabberley, Tom Maschler, Denzil and Mark Morgan, Judith Morgan, Grant Muter, Caroline Olsen, Ken Palmer, Jenny Parry, Colin Passmore, Margaret Powell, Mary Powell, Simon Powell, Eddie Procter, Mike and Shirley Rippin, Rosemary and Oliver Russell, Gill Smailes, Tony Smith, Bob Steele, Glenn Storhaug, Rita and Jack Tait, Stan Walker, Joan and Ivor Watkins and Douglas Wright.

It is dedicated to the memory of Barbara Beardsmore and of Isabel McGraghan who fostered the cohesion of the communities of Llanthony and the Grwyne Fawr by recording their past and present.

Oliver Fairclough (author, LVDHG)
and **Caroline Stanford** (editor, Landmark Trust)

Introduction

This book is about the Llanthony Valley. Also known as the
Vale of Ewyas, it is the largest of the Black Mountains valleys.
These form the eastern edge of the Brecon Beacons National
Park. The author and critic Raymond Williams (1921–1988),
who was born in nearby Pandy, compared the Black Mountains
to the fingers and thumb of one hand. To the north is Hay
Bluff, overlooking the Wye valley and the little towns of Hay
and Talgarth. From it four valleys run south-eastwards. The
eastern-most valley is the Olchon, bounded on one side by
Crib y Garth (forming the thumb). Between the Hatterall and
the Ffwddog ridges (the first two fingers) is the Vale of Ewyas
(the valley of the river Honddu), containing the hamlets of
Cwmyoy, Llanthony and Capel-y-ffin. Here the Hatterall forms
the border between England and Wales. Next, between the
Ffwddog and the Gadair Fawr, comes the Grwyne Fawr valley,
partly forested, and with a reservoir at its northern end, and
then the Grwyne Fechan, bounded by Allt Mawr. To the south
of the Black Mountains are the valley of the Usk and the towns
of Abergavenny and Crickhowell.

The Black Mountains are predominantly formed from Old
Red Sandstone laid down some 400 million years ago. Their
landscape was shaped by the ice of successive glaciations,
which gradually eroded the valleys. A great ridge south-west
of Llanvihangel Crucorney is a moraine, a mass of rock and
earth left by a melting glacier. The glaciers also cut into the

sides of the hills, causing landslips. At Cwmyoy, this left the hill Cwmyoy Graig that gives the village its name – 'the valley of the yoke'. Its church was built on the fallen rocks and earth, and has been dramatically twisted by subsidence. The Skirrid owes its jagged western edge to a similar landslip, and others almost as impressive can be seen on the eastern side of the Hatterall ridge. Even today, heavy rain can cause mudflows on the hillsides.

The lower slopes are mostly covered with well-drained brown earth soils. These are relatively fertile and reddish in colour, but the sandstone outcrops just below the surface. To the south of the river Usk the hills are mainly of carboniferous

Panoramic map of the Llanthony Valley and surrounding area, famously likened to an outstretched hand by local author Raymond Williams.

© *Contour Designs*

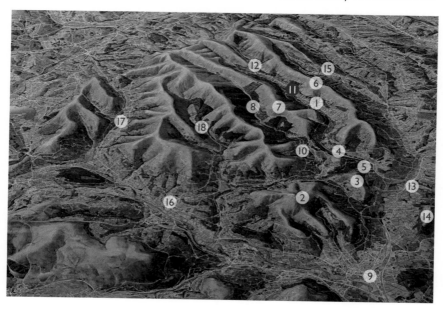

1 Llanthony Priory	7 Ffwddog	13 Llanvihangel Crucorney
2 Sugar Loaf	8 Grwyne Fawr Valley	14 Skirrid
3 Bryn Arw	9 Abergavenny	15 Olchon Valley
4 Cwmyoy	10 Partrishow	16 Crickhowell
5 Llwyn Celyn	11 **Llanthony Valley**	17 Rhiangoll Valley
6 Hatterall Ridge	12 Capel-y-ffin	18 Grwyne Fechan Valley

limestone, and in the Black Mountains burnt lime was used for centuries as a fertilizer. Higher up, the soils are often shallow, stony and acidic. The wet moorland of the high ridges is covered with peat, though this has been degraded by overgrazing, drainage and hill fires.

The Black Mountains, like the rest of Britain's uplands, are no longer in their natural state, and the grasslands of the upper slopes are home to a much-reduced range of flora and fauna. The valleys below are a patchwork of richer habitats ranging from the ffridd, where the moors meet the cultivated land, to woods and river-side meadows. Arable farming has disappeared and has been replaced by the raising of sheep and a few beef cattle on intensively managed grasslands. Sheep numbers are again at the level that existed before the outbreak of foot and mouth disease in 2001.

The early twenty-first century has seen continuing habitat loss and new diseases of trees, but also some conservation successes, with cleaner rivers and the return of the red kite and the otter. Over coming decades, the impact of climate change and Britain's withdrawal from the European Union are likely to transform the way in which the land is managed.

The human population of the Black Mountains is probably little larger today than it was in the seventeenth century, but its composition has changed greatly since the 1950s. Only a minority now have full-time work connected with the land and many small family farms have disappeared. Today's full-time residents include writers and other creatives, the owners of e-businesses, commuters and retirees.

This is one of the most special places in Britain, and tourism brings thousands of visitors. Many houses and former agricultural buildings have become second homes or holiday-lets. Nevertheless, the local community is surprisingly resilient, and welcoming to newcomers. People still know and support their neighbours, and many take part in the activities of local organisations – this book is the work of one of these, the Llanthony Valley and District History Group.

Partrishow Church (see p.40) is built on a hill above the Nant Mair stream.

I

The Valley's Story

The First People

Ten thousand years ago, after the ice sheets of the last glaciation had melted, the Black Mountains, like much of southern Britain, were heavily wooded with oaks and elms. By 6000 BC lime and ash trees were also widespread and a few Mesolithic hunter-gatherers ventured into the woods, burning off clearings to hunt deer and wild oxen. Some of their campsites survive in the archaeological record, together with scatterings of flint tools.

The first people to settle in the Black Mountains were the Neolithic farmers who began to clear the valleys of trees about six thousand years ago. Their communal burial places, stone-chambered tombs found on the lower slopes overlooking the valleys of the Wye and Usk, are the area's oldest man-made structures. Until about 3000 BC these first farmers were dependent on their grazing animals, which they moved between winter and summer pastures, rather than on cereal crops. By 2000 BC most of the upland forest had gone, peat was forming on the slopes, and people were beginning to use metal tools, first of copper, then of bronze. They were responsible for more burial sites, the cairns and round barrows that can still be seen on the ridges.

Cereals were increasingly grown, but as the area entered the Iron Age (from around 750 BC to the Roman invasion in 43 AD) the climate appears to have worsened and soil

The Iron Age hill fort of Twyn y Gaer from the air. In the middle distance (right) is Cwm Coed-y-cerrig and Bryn Arw hill, with (left) Stanton and the lower Llanthony Valley.

© Crown copyright: Royal Commission on the Ancient and Historical Monuments of Wales

fertility declined. Greater competition for resources may have followed. Twyn y Gaer at the southern end of the Ffwddog ridge, looks out towards neighbouring hill forts at Pentwyn (Bwlch Trewyn) and on the Skirrid, and was remodelled several times to provide a stone-built rampart walk and heavily defended gateway. The fort was permanently occupied, and excavation in the 1960s found evidence of iron-working there as well as the platforms of turf and wattle huts.

Celts and Romans

The Black Mountains probably formed part of the territory of the Silures, the Celtic tribal group that controlled south-east Wales. The Silures had apparently pushed eastwards in the late Iron Age. When the Romans invaded Britain in 43 AD the tribe was subdued by a network of forts built to control the river valleys. One of these was at Abergavenny, which the Romans

The Welsh Kingdoms of the early Middle Ages.

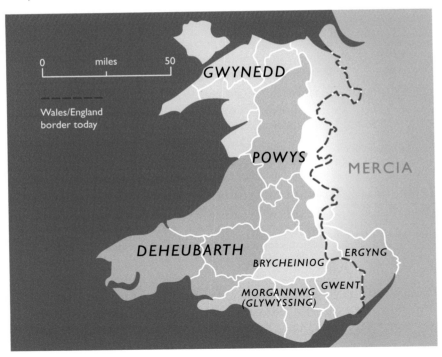

14

called Gobannium. Their conquest of Wales was completed in the mid-70s AD, but in the uplands life probably continued much as it had before the invasion.

After Roman administration collapsed in the early fifth century, Wales evolved into a number of small competing Kingdoms. To the south of the Black Mountains was the Kingdom of Gwent, comprising what became the county of Monmouthshire, and to the north and west lay Brycheiniog, settled by Irish raiders from west Wales. The people of Gwent were also threatened by the Saxons, especially in the eastern district or commote of Ergyng (now south Herefordshire, west of the Wye), which came under English control in about 1000 AD, while remaining Welsh in culture. This was therefore a frontier zone, and it was to remain a borderland for almost a thousand years. Another constant from the late Roman period was Christianity. Several churches in north Monmouthshire were founded in the seventh century, and St Michael's at Llanvihangel Crucorney is recorded in a charter of about 970 AD.

Marcher Lordships

The frontier between the English and the Welsh had been relatively stable since the 700s, but late in the eleventh century, soon after their conquest of England, Norman barons

Hay-on-Wye

HAY

Wye

Wye

Hereford

ENGLAND

Monnow

EWYAS

Llanthony
Priory

Cwmyoy

Grosmont

Wye

Monnow

Blaenllynfi

Partrishow

THREE
CASTLES

Usk

Tretower

Llanvihangel
Crucorney

Crickhowell

Abergavenny

Monmouth

BLAENLLYNFI

ABERGAVENNY

MONMOUTH

Forest
of
Dean

Rhymni

Ebbw

Usk

Usk

Wye

WALES

USK

Chepstow

Newport

0   miles   10

- - - - - -
Wales/England
border today

SEVERN

The Lordships
of south-east
Wales during the
thirteenth century.

holding lands on the border began to advance into Wales, not only to raid, but to conquer. Their conquests became a frontier region outside royal authority, known as the March.

The Black Mountains west of the Grwyne Fawr River lay in the commote of Ystrad Yw in Brycheiniog, which fell to Bernard de Neufmarché, the founder, in 1093, of the Lordships of Brecon and Blaenllynfi. The Honddu and Olchon valleys became part of the Lordship of Ewyas, held by the de Lacys. However these conquests were insecure, as other Welsh lordships survived in South Wales, and the still-independent Princes of Deheubarth and Gwynedd threatened and sometimes dominated the March (the border area under Norman control) until well into the thirteenth century.

In 1136 Richard Fitz Gilbert de Clare, the Norman Lord of Chepstow and Ceredigion, was killed in the Grwyne Fawr by the soldiers of Morgan ab Owain of Caerleon, a shocking event recalled by Gerald of Wales who passed that way in 1188. In 1175 William de Broase invited Seisyll ap Dyfnwal and his followers to Abergavenny Castle as his guests and massacred them there. The Welsh captured and burnt the place in 1182, and Abergavenny was destroyed again in 1233. This strategic stalemate in the March ended only with Edward I's conquest of Gwynedd in 1282–3.

Many of the Marcher Lords founded monasteries, and in the 1180s Hugh de Lacy of Ewyas further endowed the struggling Augustinian Priory of Llanthony Prima. In the mid-twelfth century this had become a target for Welsh raids, and its canons retreated to Gloucester. Only in the 1180s did they return to start work on the church we see today, separating formally from their daughter house in Gloucester, Llanthony Secunda, in 1205.

The climate of northern Europe worsened in the early fourteenth century, causing famine in Wales in the years around 1320. In 1349 the great plague known as the Black Death reached the March, killing perhaps a third of the population. It was to return in 1361 and 1369, and rural life was profoundly affected by a shortage of labour, with land left uncultivated and 'in decay'.

Owain Glyn Dŵr's Revolt and its Aftermath

Edward I's conquest left the Welsh subject to legal, economic and linguistic discrimination, and following the deposition of Richard II by his cousin Henry IV in 1399, a new rebellion began against English rule. This coalesced around the inspirational Owain ap Gruffydd Fychan, Lord of Glyndyfrwy, later known as Owain Glyn Dŵr. In May 1401 the tenants of Abergavenny Lordship rose against their lord and following Owain's crushing defeat of an army led by the gentry of Herefordshire and the Southern March in a battle near Knighton in June 1402, English administration in the area collapsed. The towns with their mostly English inhabitants were attacked. Abergavenny and Hay were torched, though the castles held out. Brecon was besieged while Crickhowell was taken and its castle sacked. The Welsh also crossed into Herefordshire where they 'burned houses, killed people, took prisoners, and ravaged the land'. The English won battles at nearby Grosmont and Usk in 1405, but later that year a French army, allied to Glyn Dŵr, campaigned in the March.

The revolt was partly a civil war, but Glyn Dŵr, who declared himself Prince of Wales, attracted strong popular support. Only in 1407 did war turn in favour of the English Crown, and the rebellion smouldered on for several more years.

Destruction in the area seems to have been massive. Llanthony Priory received a payment from the Crown in 1405 in recognition of its losses. Much weakened, it merged with Llanthony Secunda in Gloucester in 1481. The war left a legacy of lawlessness, especially cross-border banditry and cattle-raiding. During the Wars of the Roses (1459–1485), supporters of both sides recruited locally. It was probably only when Henry VII's grip on power became secure at the end of the fifteenth century that the Llanthony Valley recovered the level of population and economic activity it had enjoyed two centuries earlier. His son Henry VIII was to deploy the power of the Tudor state to pacify the March.

The Sixteenth Century:
The Acts of Union and the Reformation

Henry VIII's Acts of Union of 1536 and 1543 abolished the Marcher lordships and incorporated them into seven new counties. English also became the language of the law and of public office. Most of the Black Mountains became part of the new counties of Monmouthshire and Brecknockshire. However the Olchon and the Ffwddog were added to Herefordshire.

Following his break with Rome, Henry dissolved the monasteries of England and Wales, and in 1538 Llanthony Priory was surrendered to his officials by its prior and four remaining canons. Its lands were leased to Nicholas Arnold (c.1509–1580),

The counties of
Wales from 1536.

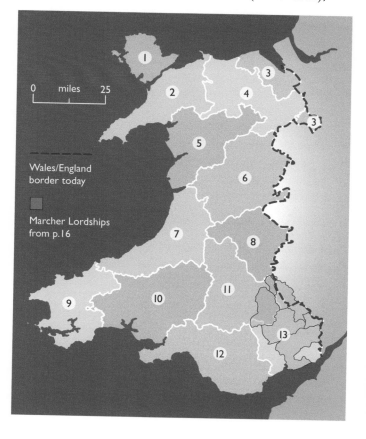

0   miles   25

Wales/England
border today

Marcher Lordships
from p.16

1 Anglesey
2 Caernarvonshire
3 Flintshire
4 Denbighshire
5 Merioneth
6 Montgomeryshire
7 Cardiganshire
8 Radnorshire
9 Pembrokeshire
10 Carmarthenshire
11 Brecknockshire
12 Glamorgan
13 Monmouthshire

a former tenant of the priory, who purchased the site, together with Cwmyoy manor and some of its other estates in 1546.

Arnold left Llanthony to his younger son John Arnold (d. 1606). His grandson and namesake Nicholas Arnold (1599–c.1665) extended his family's control of the valley, purchasing the manor of Llanvihangel Crucorney in 1627. The Arnolds struggled to make the estate pay. Much of the land was held on lengthy copyhold tenancies at low rents. It was therefore the tenants who benefited from the growing prosperity enjoyed by farmers in the late sixteenth and early seventeenth centuries. From the wills of these people, it is clear that much of their wealth was in the form of cattle.

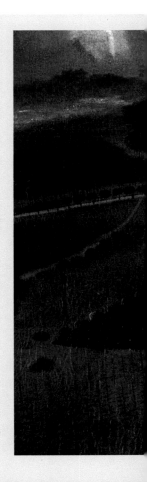

JOHN ARNOLD (1635–1702)

Between 1678 and 1681 Britain was gripped with anti-Catholic hysteria whipped up by the fictitious 'Popish Plot' to assassinate Charles II. John Arnold became famous in March 1678 when he gave lurid 'evidence' to the House of Commons about Catholic activities in Monmouthshire. He was probably motivated by personal animosity towards the Lord Lieutenant, the Marquess of Worcester, whom he accused of Catholic sympathies. He then led an onslaught on the Catholic priests ministering in the area. Several were arrested, including Father David Lewis (1616–79), who was executed at Usk (canonized a saint in 1970).

In April 1680 an alleged Catholic attempt on Arnold's life made him a national hero, and he was elected MP for Monmouthshire. However a scandalous accusation that Worcester was involved in the Popish Plot saw him imprisoned from 1683 to 1686. He was re-elected to Parliament in 1689 after the flight of the Catholic King James II, where he 'discovered' more Catholic plots. Though he sat until 1698, he never obtained the lucrative government office he needed, and his extreme Whig and Protestant views made him deeply unpopular.

The Seventeenth and Eighteenth Centuries

From around 1600 to 1640, the population grew, and the reshaping of the landscape continued as new farms were built on higher ground. In contrast, the owner of the Llanthony estate, Nicholas Arnold, was in deep financial trouble and relations with his tenants deteriorated. He was to spend the last twenty years of his life in a debtors' prison, with the estate in the hands of his creditors.

Monmouthshire was held by the Royalists for much of the Civil War, but saw heavy fighting in 1645–6. John Arnold (1635–1702), who succeeded his father in 1665, was a fanatical Protestant and spent much of his life hunting

Llanvihangel Court, John Arnold's home, painted in about 1688. Private Collection.

down Catholic sympathisers. His son, another
Nicholas Arnold, inherited a neglected and
impoverished estate, and was taken to court by
his tenants who sought to establish the customs
of the manor in their favour. Nicholas won,
but the case dragged on for years, and when
he died childless his sisters sold Llanthony
and Llanvihangel Crucorney in 1726 to
Edward Harley (1664–1735) of Eywood in
Herefordshire, and brother of the 1st Earl of
Oxford, for £36,186-11s.

All the major landowners in the area were
now absentees, and even the freehold farms
were increasingly occupied by tenants. The
population probably peaked around the middle
of the eighteenth century. The poor then began
to leave, some to find work in the developing
ironworks, quarries and mines around Blaenavon.
These nearby industrial communities provided
a new market for produce and a long process of
farm consolidation into larger units began.

The Llanthony estate changed hands
twice, being sold to the Indian nabob Sir Mark
Wood (1750–1829) in 1799, and then to
Walter Savage Landor in 1807. Llanvihangel
Crucorney was sold separately by the Harleys
in 1801. The upper part of the Grwyne Fawr,
which had belonged to the Williams family
of Gwernyfed, near Talgarth, passed to the
London socialite John Macnamara in 1796, and
was sold in 1847 to the ironmaster Sir Joseph
Bailey (1783–1858).

Llanvihangel Court, bought by the Arnold family
in 1627, and substantially enlarged in the early
seventeenth century. Photographed in 1924.

© *Country Life*

WALTER SAVAGE LANDOR (1775–1864)

A poet and essayist, Landor had made his name with a verse romance *Gebir*, which was admired by Wordsworth and Shelley. Quarrelsome, headstrong and romantic, in 1807 he fell in love with Llanthony and sold properties elsewhere to buy the estate. He began to build a new house at the Siarpal farm, planting thousands of sweet chestnuts, cedars and larches, and introducing Spanish merino sheep. In a letter of 1812, Landor described the pleasures of country life, writing:

> *Homeward I turn; o'er Hatterils rocks*
> *I see my trees, I hear my flocks.*
> *Where alders mourned their fruitless bed*
> *Ten thousand cedars raise their head.*
> *And from Segovia's hills remote*
> *My sheep enrich my neighbour's cote*

The ruins of the Siarpal. Landor lived briefly in part of his new house, which was largely demolished after he left the valley.

© Simon Powell

Walter Savage
Landor, painted in
middle age by William
Fisher, 1839.

© *National Portrait
Gallery, London*

Landor's attempts to manage the estate were disastrous.
He quarreled with everyone from the Lord Lieutenant to
his own tenants, declaring 'drunkenness, idleness, mischief
and revenge' to be the principal characteristics of the
Welsh. Harassed by lawsuits and debts, he left Britain in
1814 and spent much of the rest of his life in Italy.

The Nineteenth Century: Prosperity and Depression

The Ordnance Survey published the first accurate map of the Black Mountains in 1830. From this, and from the maps made in the 1840s as part of the process of establishing cash payments to the Church instead of tithes, we know a good deal about the landscape of the valleys in the nineteenth century. Local newspapers, church and chapel registers, family stories, even the memories recorded in the twentieth century, also begin to throw light on people's everyday lives.

Meanwhile Abergavenny trebled in size during the nineteenth century. The Monmouthshire and Brecon canal was completed in 1812, and the Newport, Abergavenny and Hereford railway, which opened in 1854, connected the area to the wider economy of south-east Wales.

Then, from the 1870s, farm prices tumbled with the arrival of cheaper food from the Americas and Australasia. Marginal land was abandoned, and depopulation increased. Cultural change was similarly massive. Religious non-conformity grew rapidly, and elementary education, conducted in English, became compulsory in 1880.

The former railway station at Llanvihangel Crucorney, on the edge of the village. The station at Pandy, only a couple of miles down the track towards Hereford, was also closed in 1958.

© Simon Powell

The Twentieth Century: War, Renewal, and Tourism

The Agricultural Depression that hit the region in the late-nineteenth century continued to devastate living standards until the Second World War. The agricultural workforce, conscripted during the First World War, dwindled further in size after it, and more land was abandoned. As coal and later oil became available for heating, the woodlands were no longer managed for fuel. Mechanisation was limited by the terrain, and by the 1930s the traditional, self-sufficient farming community was close to collapse. It was saved by the outbreak of war in 1939, which brought State control of agriculture to increase food production, and some government support continued after 1945.

Sgt Guy Powell, Cwm Farm, Cwmyoy, about 1940. Many of the farmers in the Valley also served in the Home Guard during World War II.

Twentieth-century amenities slowly arrived. In 1912 the Abertillery Water Board began work on an eighty-hectare (200-acre) reservoir at the head of the Grwyne Fawr (see p.70). This required the building of a new road, and temporary homes for the workforce. Other roads were metalled, and a daily bus service to Llanthony came in 1930 – and went in 1990. The ownership of motor vehicles – often small vans – first became common during the 1960s. Electricity from the National Grid reached the Llanthony Valley in the early 1960s, together with a telephone service. Only then were people no longer dependent on oil lamps and candles for lighting, or coal-fired ranges for cooking, and could refrigerate food. For a while both Cwmyoy and Forest Coalpit had a general store. Houses were gradually improved. Outside privies gave way to flushing toilets and septic tanks. Central heating and bathrooms were installed, and some social housing built.

Capel-y-ffin in about 1950, when the upper valley road was still a track. The Anglican chapel is in the background.

© Christine Olsen

Above Canadian soldiers felling trees in the Llanthony Valley during World War One.

© *Knight family*

Below The James family of Trefeddw at the Pandy show in 1943

© *Edith Davies*

The mobile shop, later permanently parked in Cwmyoy.

Blackwell and Son, later part of Red and White Coaches, operated a bus service between Llanthony and Abergavenny from 1930.

When the Llanthony estate was sold between the late '50s and 1966 many of the tenants purchased their holdings. There was further farm amalgamation, and heavier stocking, as well as an expansion of commercial forestry. The Brecon Beacons National Park was established to safeguard the landscape in 1957, and a Youth Hostel opened at Capel-y-ffin a year later. Even so, more houses were abandoned or condemned and in 1960 the BBC Welsh Home Service made a documentary about Llanthony called 'The Lonely Valley', portraying it as an introspective place with a brooding self-sufficiency, to locals' irritation. Then, in the 1960s second-home owners began to buy and restore often-derelict properties.

By the 1980s, environmental and historical conservation had entered public consciousness, and grants to the farming community became set according to environmental practice as well as production levels. Tourism, and leisure – hill walking, pony-trekking and cycling – became a mainstay of the local economy alongside farming and forestry.

Pony-trekking near Llanthony in 2004.

2

Community and Culture

A Welsh Society

Although the Black Mountains straddle today's border between England and Wales, their people were largely Welsh in language and culture until the mid-nineteenth century. At the end of the Middle Ages most people in the upland areas of the March were the descendants of the Welsh kinship groups that had occupied the land before the Normans. Despite an English ban on Welshmen holding public offices, their leaders were often stewards of the by now absentee lords. Regarded as gentry or uchelwyr by their neighbours, some claimed descent from the Welsh princes and built up landed estates of their own.

After the Union with England in 1536 these people became the ruling elite of the new counties. The manor of Llanvihangel Crucorney passed through several such families between the 1450s and its sale in the early-seventeenth century to Nicholas Arnold. Pride in family was strong, and the kinsmen of the greater landowners were also regarded as gentry, even if they were little wealthier than their neighbours. The fortunes of these 'parish gentry' and other freeholders were improving during the sixteenth century. They held their land by inheritance, and had the right to graze stock on the upland commons. The copyhold tenants, who were numerous in Cwmyoy, had long leases at low rents.

This social structure dictated how people lived. There are few villages in the Black Mountains, and nearly everyone lived

Shearing, Llanthony, c.1937.

Gwent Archives

33

on the farms that dotted the landscape. These were all small in size by today's standards, and the community was headed by the few fortunate men who farmed as much as fifty hectares (120 acres). The specialist craftsmen – in the seventeenth and eighteenth centuries these included blacksmiths, millers, coopers, shoemakers, tailors, carpenters and masons – were also usually part-time farmers.

Despite this pattern of isolated settlement, life before the twentieth century was partly communal, as neighbours had to come together for work that needed many hands, such as gathering of stock on the mountains, shearing or harvest, and supported each other when ill. People, especially men, often

The Skirrid (Ysgyryd Fawr) said to have been shattered during the Crucifixion.

© Simon Powell

married late – and even in the nineteenth century it was not uncommon for children to be born outside marriage. Farm labourers and maid servants usually lodged in the house of their master or mistress, until they had enough money to marry and take on a farm of their own. Premature death and remarriage were frequent.

Welsh remained the language of everyday life. In the sixteenth century farms in the valley began to be given descriptive names. These are entirely Welsh, as are nearly all the field names recorded in the tithe apportionment surveys of the 1840s. The Welsh spoken was the speech of Gwent, known as *Gwenhwyseg*, which survives only in the local pronunciation of place names. English was the language of public life, business and the law. Bilingualism spread gradually but as late as the eighteenth century was probably still confined to the wealthier farmers, who served as parish constables or churchwardens.

Belief in the Middle Ages

Religion permeated every aspect of life. The churches of Llanvihangel Crucorney, Cwmyoy, Partrishow, Cloddock and Llanveynoe are all pre-Norman in origin. St David, the sixth-century patron saint of Wales, was said to have built a chapel at Llanthony (its name is a corruption of Llan Dewi Nant Honddu – the church of St David by the Honddu). Rediscovered by William de Lacy in the 1090s this became the site of the Augustinian Priory of Llanthony.

The Skirrid, known as the Holy Mountain, was said to have been struck by lightning at the moment of Christ's crucifixion. The now-lost chapel of St Michael on its summit was a place of pilgrimage. As the parishes are mostly very large, there were also medieval chapels 'of ease' at Capel-y-ffin and Stanton, allowing mass and prayer closer to home. Although monastic life at the Priory was at a low ebb by the early sixteenth century, the parish churches were flourishing. Partrishow still has an astonishingly beautiful rood screen, installed in the church around 1500.

LLANTHONY PRIORY

Although the small community at Llanthony became Augustinian canons (priests living as monks but carrying out pastoral work) in about 1120, what remains of their church was built between about 1180 and 1220, just as the Norman style with its round arches and massive columns was giving way to Early English Gothic, characterised by pointed arches, lancet windows and ribbed stone vaults.

Like most large medieval churches the Priory is cross-shaped with a central crossing tower and two smaller towers at the west end.

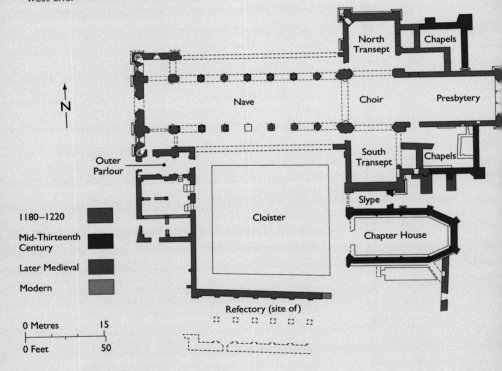

Ground plan of Llanthony Priory.

© Crown copyright (2017) Cadw, Welsh Government

The south transept, and crossing tower, the upper part of which was taken down in 1808.

© Simon Powell

When the Priory was dissolved in 1538, lead was removed from
the church roof, and it was left to decay. The Prior's lodging and
part of the west range of the cloister became a house (later
Court Farm and the Abbey Hotel).

By 1732, the priory church was a ruin, and more of the building
fell down over the next century, after which attempts began to
save what remained. From the 1770s it was much visited by early
tourists in search of the Picturesque (see p. 97).

The towers of the west
front were originally
taller and the three
lancet windows of the
nave collapsed in 1803.

© *Simon Powell*

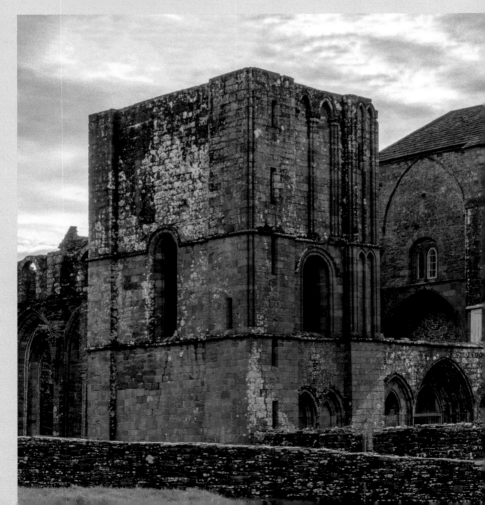

The Abbey Hotel, built around 1800 and incorporating
one of the towers of the church's west front.

© Simon Powell

Above The monastery's infirmary and its chapel were
converted into a Protestant church.

© Simon Powell

PARTRISHOW CHURCH

This remote little church is dedicated to
St Issui, or Ishow, a sixth-century hermit
who lived near a holy well (actually a
shallow spring beside the Nant Mair
stream). Following his murder, his cell
became a place of pilgrimage, and a
church, containing his grave or shrine,
was later built on the hillside nearby. This
has changed little since the sixteenth
century, though the south wall was close
to collapse in 1908, when the building was
sympathetically restored by W.D. Caroe
(1857–1938).

The oldest part of this church is its font,
which dates from about 1060–1100.

Right Memorial to William Price (1718–1793),
his wife and one of his daughters carved by
John Brute (1752–1834). The Price family lived
at Fforddlas for a century from the 1730s.

© Simon Powell

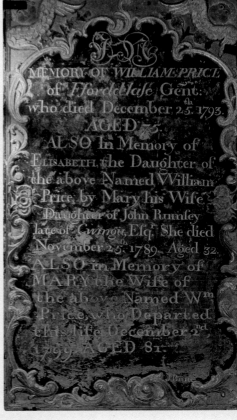

Left The font's inscription reads
'MENHIR ME FECIT I(N) TE(M)
PORE GENILLIN' ('Menhir made
me in the time of Genillin'). Genillin
or Cynhillin was Lord of Ystrad Yw
in the late eleventh century.

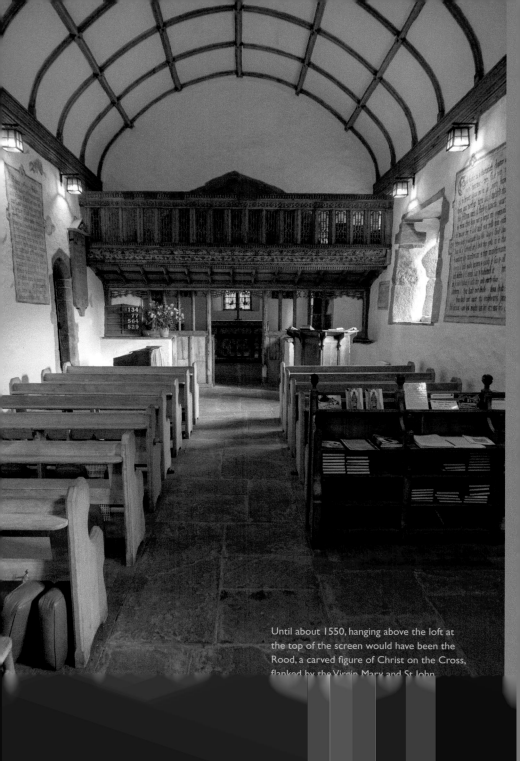

Until about 1550, hanging above the loft at
the top of the screen would have been the
Rood, a carved figure of Christ on the Cross,
flanked by the Virgin Mary and St John.

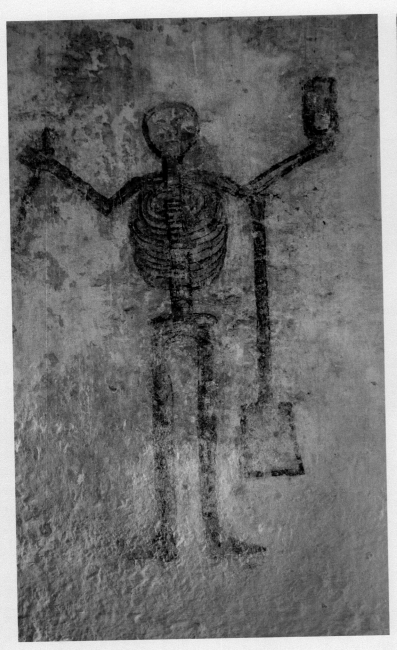

Left A seventeenth century 'Doom' figure, a skeleton bearing an hour gla scythe and spade, a reminder that in tir all must be cut dow and die.

© Simon Powell

The parapet is made up of seventeen openwork tracery panels, carried on a great horizontal beam or bressumer carved with three trails of running ornament.

© Simon Powell

The early sixteenth-century oak rood screen, separating the nave from the chancel, is one of the glories of this church. The antiquarian Richard Fenton, who came here at the beginning of the nineteenth century, described it as 'the most perfect and elegant Rood loft now standing in the Kingdom'.

After the Reformation the walls were painted with texts from the Bible and Book of Common Prayer. The chancel is full of exuberant carved and painted eighteenth-century memorial tablets in painted sandstone. Many of these were made by the Brute family of masons from Llanbedr, and they commemorate the farming families of the Grwyne Fawr.

The Reformation

The break with Rome in 1535 brought little immediate change. Llanthony Priory was suppressed in 1538, and its infirmary was converted into a church for the people of the upper part of the valley. In Cwmyoy Church a statue of St Leonard of Noblac, credited with miracles for women in labour and diseases of cattle, was destroyed as an idolatrous image. 1548 saw the replacement of the Latin liturgy by the English Book of Common Prayer. The latter was translated into Welsh in 1567, and the Bible followed in 1588. Both were used in the valley's churches for centuries. In 1800, church services at Cwmyoy and Llanthony were still conducted alternately in Welsh and English.

Catholics and Nonconformists

Catholic beliefs did not disappear entirely, and in the late sixteenth century a few foreign-trained priests arrived to keep their faith alive in Monmouthshire. Soon after, a secret Jesuit school was established at the Cwm, near Llanrothal on the Herefordshire border. Claiming that the area was full of Papists, John Arnold ran a violently anti-Catholic campaign in the 1670s (see p.20). Despite this persecution, some farming families, especially their women, took part in an isolated and distinctively Welsh Catholic culture throughout the seventeenth century. There were thirty-six recusants (known Catholics) in Cwmyoy in 1706, and about half that number fifty years later.

It was the valley's Catholic past that later attracted the eccentric mystic Joseph Leycester Lyne (1837–1908), known as Father Ignatius. His life's goal was to introduce his own eclectic version of monasticism into the Anglican Church. In 1869 Ignatius bought land at Capel-y-ffin, where he began to build a replica of Llanthony's priory church. Of this only the choir was built,

together with accommodation for Ignatius's handful of
monks. Ignatius was an inspirational figure, passionate about
everything from the Welsh language to the Flat Earth theory.
However the community dwindled, and after his death
the monastery passed to the Catholic monks of Caldey in
Pembrokeshire, who leased the derelict buildings to the artist
Eric Gill (1882–1940).

Protestant Nonconformity also has deep roots in the
Black Mountains. There was a congregation of Baptists in the
Olchon valley from the 1630s. Like the Catholics, they were
frequently persecuted by the authorities until the Toleration
Act of 1689, and held meetings in secret. The small Baptist
chapel in Capel-y-ffin was built in 1762 by David and William
Prosser as a meeting place for the Baptist congregation there
and in the Olchon. This had fifty-two members in 1800,
and Nonconformity grew rapidly during the nineteenth
century. Though the more prosperous farming families mostly
continued to attend church, much of the community turned to
the chapels.

In Cwmyoy parish in 1851, the congregations of the
Tabernacle Baptist Chapel, Ffwddog (built 1838), the
Bethlehem Calvinistic Methodist Chapel (built 1839) and
Henllan Baptist Chapel (built 1840) numbered over three
hundred, twice that of St Martin's Church and St David's at
Llanthony. A key Baptist belief was adult baptism and all their
chapels are by the Honddu or Grwyne Fawr rivers.

Schools and Linguistic Change

The chapels had Sunday Schools for both children and adults
where the teaching, like the meetings themselves, was
mostly in Welsh. There were few opportunities for education
elsewhere. Neither a school for poor girls at Partrishow
endowed in 1725 nor a mid-eighteenth-century church school
in Cwmyoy lasted long.

In the 1850s an elderly tailor Lewis Williams ran a school
with his wife Margaret in their Partrishow cottage, and by then
a network of elementary schools was developing. A National

The Baptist Chapel
at Capel-y-ffin, the
oldest in the valley, is
built in whitewashed
sandstone, with simple
pointed windows.

© Simon Powell

(Anglican Church) School opened in Llanthony – initially in Maes-y-Beran farmhouse and then in a new building at the west end of St David's Church. In 1855 a one-room school with a teacher's house was built in Cwmyoy for the lower part of the valley. Soon afterwards the State began to contribute to the cost of elementary education, and the 1870 Education Act created a network of school management boards, funded from the local rates. Cwmyoy School became a non-denominational board school in 1878, while from the late 1860s children from the Grwyne Fawr went to a school in Forest Coalpit.

Education up to the age of ten became compulsory in 1880 (parents had to pay 2d or 3d a week until 1891) but attendance was often poor at busy times during the farming year, or when the weather was bad. The children were taught reading and writing in English. Consequently, first they, and then the wider community, rapidly lost their Welsh. A commentator writing in 1892 noted that 'at Cwmyoy Welsh is nearly extinct' while in Llanthony only those in their fifties could speak Welsh.

Tabernacle Baptist Chapel, Ffwddog, built in 1837 by Samuel and James Thomas, father and son from nearby New Inn Farm.

© *Simon Powell*

Plans for Cwmyoy School, drawn by the
Abergavenny architect J. H. Evins in 1855.

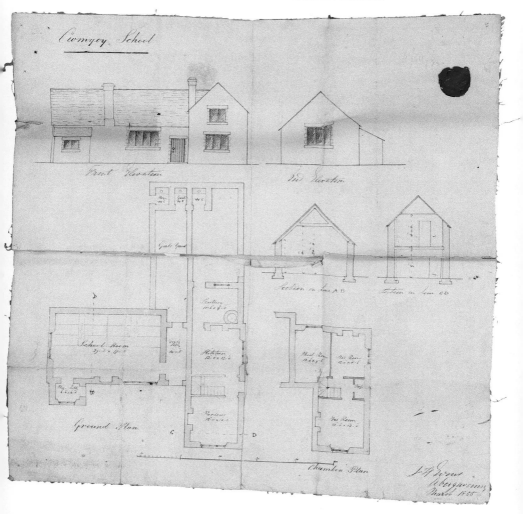

Children and staff of Llanthony School (background on right) playing 'The Farmer's in his Den', early 1950s.

© Christine Olsen

Off to Barry Island. A Henllan Chapel outing, late 1950s.

© Jenny Francis

Local Government Reform

The new schools were mostly established and governed by
the wealthier farmers. These, under the loose supervision of
the justices of the peace, had for generations been the unpaid
leaders of their communities, serving annual terms as petty
constables (elected at the manor courts that continued to
function until the late-eighteenth century), as church wardens
and as overseers of the poor, taking responsibility locally for
the destitute. From 1837 Cwmyoy was one of the parishes
in the Abergavenny Union and sent its paupers to the town's
workhouse. In 1857 the petty constables' responsibilities
passed to the newly-established county police forces. A
measure of local democracy arrived in 1888 with the creation
of directly elected county councils throughout England and
Wales, and in 1891 the Ffwddog ridge was transferred from
Herefordshire to Monmouthshire. Rural district councils for
Abergavenny, Crickhowell and Hay, responsible for most local
services, followed in 1894.

The Community in the Twentieth Century

The population of the valleys was in decline throughout the
nineteenth century, but attendance at church or chapel was
almost universal, and nobody worked on Sundays. In the
years before 1914, both Cwmyoy and Partrishow churches
underwent extensive restoration. In 1904–5 the so-called Great
Revival of Christian belief swept across South Wales, bringing
new conversions, especially to Nonconformity.

After the First World War the chapels were still full and,
other than weekly trips to Abergavenny market, 'provided the
only social life the valley knew' until well into the twentieth
century. People were either 'Church' or 'Chapel', a difference
that sometimes divided neighbours and families. Both
organised occasional social events, such as tea parties, concerts
and summer trips to Barry Island.

Probably the first non-denominational organisation was
the Cwmyoy branch of the Women's Institute, established in
1926. Men who resisted the Chapels' Temperance campaigns

Above The Globe, Forest Coalpit, a pub which closed in 1916.

Left Cwmyoy Memorial Hall, the gift of Mrs Constance Molyneux in 1928.

© *Simon Powell*

had only the valley's pubs in which to socialise. These in 1900 were more numerous than today and included the Black Lion in Cwmyoy village, the New Inn, Ffwddog, and the Globe at Forest Coalpit. Quoits was a popular male sport. In 1928 Constance Molyneux built Cwmyoy's community hall as a memorial to her husband. Remembered for dances in the '30s and '40s, this has been heavily used by local people ever since. Although some children went on to secondary schools in Abergavenny, the Valley remained an inward-looking farming community into the 1960s. Today, all the primary schools in the valley have closed, and despite some inward immigration the area faces the challenges of isolation common in remote rural areas.

CONSTANCE MOLYNEUX (1892–1981)

Born in Bridgend, Constance Hood married William
Molyneux in 1924. Together they bought one of the
few big houses in the area, Trewyn, and a few fields
around it for £4,000. Her husband died in 1928 at
the age of fifty, but 'Mrs Moly' lived on at Trewyn for
over fifty years, with a household that dwindled from
a cook, four maids and a chauffeur to a single daily
woman. A champion of the Church, the Girl Guides
and the Women's Institute, she served as a magistrate
into her seventies. She was respected, and even loved,
as a leader of the Valley community.

Constance Molyneux (with dog) and the
Cwmyoy Women's Institute in about 1960.

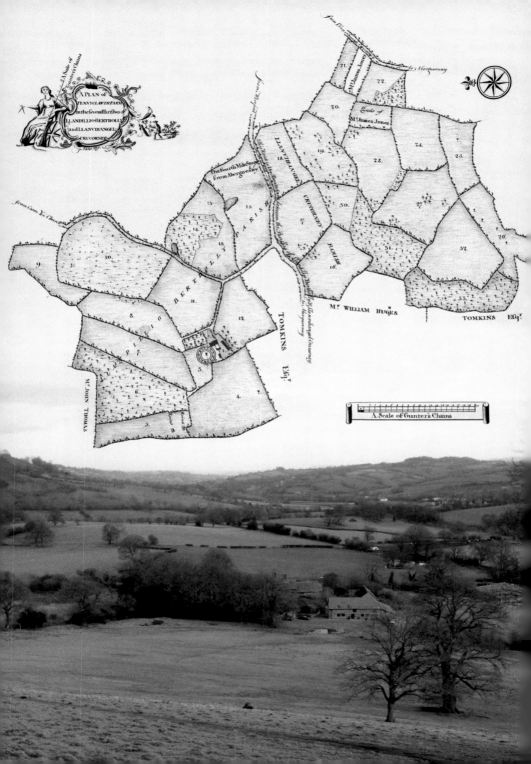

3

Working the Land

The Black Mountains have been the home of farming communities for at least five thousand years. What we see today is a countryside largely formed in the fifteenth century, in the aftermath of the Black Death and of the Glyn Dŵr rising. Although only a handful of the valley's farmhouses pre-date 1500, many others occupy likely medieval sites, often at a height of around 250 metres where springs are common. Farms have fluctuated in size over the centuries as land was bought, sold, inherited or rented, and many smaller ones have since been abandoned. Nevertheless, the upland farms and those in the valley bottom all worked the land in much the same way. In order both to feed the community and to produce a cash surplus, they were 'mixed' farms – growing cereals as well as rearing cattle and sheep.

Cattle, Sheep and Corn

The fields on either side of the rivers Olchon, Honddu and Grwyne Fawr are relatively flat and fertile, but the soil is heavy and sometimes waterlogged. These fields were mostly meadows, cut two or three times a year for hay. Today much of the Black Mountains is sheep pasture, but in the 1840s nearly half the acreage of many farms was arable. The lower slopes, increasingly divided up by walls or hedges into enclosed fields from the sixteenth century, were a mix of ploughed land and pasture, often with a belt of woodland above.

A lowland farm: Penyclawdd Court, near Llanvihangel Crucorney, surveyed in 1775, showing a mix of arable and pasture. The view today (**below**) looking from field 4.

Gwent Archives

An upland farm: Ty-hwnt-y-bwlch, Cwmyoy, with enclosed fields giving way to moorland waste commons beyond the line of the mountain wall.

© Simon Powell

The open hillsides above about 350 metres (1200 feet) provided many hectares of additional grazing, and were also a source of peat (burnt on the fire) and bracken (used for bedding animals during the winter). Before 1900 few farms were more than fifty hectares (125 acres) in extent, and most were a lot less. In the Llanthony Valley and the side cwms there was sometimes a second farm above the first. These may have originated as hafodydd (summer dwellings) which were first permanently settled in the fifteenth century.

The wills made by some of the inhabitants between the mid-1500s and early-1800s, and especially the accompanying inventories of their possessions, tell us a good deal about how the land was farmed. In the sixteenth century cattle raising was the principal source of wealth, and some herds were quite large. Watkin Winstonne of Cwmyoy, who died in 1584, had at least thirty-eight beasts, heifers, milk cows and bullocks, as well as two oxen to pull his plough, and nine horses.

Opposite A ploughing competition below Bryn Arw, about 1950.

© *Edith Davies*

Most of these cattle were sold as adult beasts for meat, and may have been taken by drovers from West Wales as far as London. Cattle require feeding during the winter, and hay from the meadows was supplemented by oats, and by makeshifts such as chopped holly leaves. They also need shelter, though only the more valuable animals seem to have been wintered under cover.

Godfrey Williams with a Hereford bull, Court Farm, Llanthony, winter 1962–3.

© *Jenny Francis*

During the seventeenth century there was a move towards selling bullocks young for fattening elsewhere, and sheep became increasingly important, for wool, meat and milk. Flocks were still quite small, averaging about forty ewes, but had grown considerably by 1800 (John Powell had over four hundred sheep at Ty'n-y-llwyn, Partrishow, in 1807).

Despite the altitude, the high rainfall, and the acidic soil, a wide range of cereals was grown. The Partrishow yeoman

Above left A team of horses at Penywern, Cwmyoy, about 1938.
© *Jenny Francis*

Above right Harvesting corn with a horse-drawn mower and tying the sheaves by hand, Penywern, 1939.
© *Jenny Francis*

Cecillt Watkin who died in 1666 grew barley, wheat, rye and oats. He also owned ten cattle, as well as forty-four sheep and forty lambs. Fields were ploughed using teams of two, four or six oxen. This was slow work – it could take several weeks to plough twenty hectares – but oxen could cope with the steep ground, and were only replaced by the new breeds of heavy horses in the nineteenth century. Oxen were valuable animals, worth around £5 each in the mid-eighteenth century, or about twice as much as a horse.

Above Bert Pugh taking a barrel of cider on a horse-drawn
sledge to the haymakers, Llwyncelyn, Grwyne Fawr, about 1930.

© *Stan Walker*

Below Raking hay, probably at
Fforddlas, Grwyne Fawr, about 1930.

© *Stan Walker*

The inventories list both corn in the barn (harvested and threshed grain), and growing corn in the ground. Barley seems to have been the principal crop, though there were eight hectares (twenty acres) of wheat growing at Llwyn Celyn, Cwmyoy, in 1848. Corn and hay were moved over rough ground on a gambo, a low two-wheeled cart, or on a wheel-less sledge. Every inventory mentions 'implements of husbandry' with values ranging from a few shillings to £25 or £30 in the early nineteenth century on the larger farms. These included ploughs, harrows, and a range of hand tools.

There were several small water mills on both the Honddu and the Grwyne Fawr in the eighteenth century, and the mills at Llanvihangel Crucorney and Pontyspig continued to grind flour into the twentieth century. Much of this flour was for local use – nearly every house had a bread oven (women usually baked once a week), and some of the larger farmsteads included a separate bake house. Barley was also malted to make beer, and many farms had cider orchards. Pigs provided some fresh meat and bacon. Most households kept hens, ducks and geese, grew vegetables and fruit, and made honey.

Mary Ellen Pugh (later James) feeding her hens at Llwyncelyn, Grwyne Fawr, about 1930.

© Stan Walker

The Work of Women

Women were responsible for the house and family, and also helped in the fields. They did the dairying, making sheep's cheese for home consumption, and using cow's milk for the butter and cheese they took to Abergavenny market together with rabbits, poultry, eggs, and anything else they could sell for cash. They also made some of the family's clothes. In the early twentieth century their work was often arranged round a strict weekly timetable – Monday would be washing, Tuesday would be market, Wednesday would be baking, Friday was cleaning. Their daughters would assist as soon as they were old enough, and all but the smallest farms had a teenage girl helper or maid. However many girls left the area to work as indoor servants elsewhere.

Tenants and Landlords

Most farmers were tenants, as much of the Vale of Ewyas belonged to the Llanthony estate until the mid-twentieth century. Gradually long leases for three or four lives at fixed rents, supplemented by large payments when leases were renewed, gave way to annual agreements. This process was still incomplete when the Llanthony estate was sold in 1799.

Opposite Mary Drew (later Williams) haymaking with Violet and Mary James at Trefeddw, about 1923.

© Edith Davies

Right The mountain wall, built in sections, probably in the seventeenth century, and running for nearly fifty miles, marks the boundary between the upland commons and the enclosed fields.

© Arthur James

Court Farm, Llanthony (105 hectares), was then held on an annual lease at £70, while Ty-hwnt-y-bwlch, Cwmyoy (just eighteen hectares) was sublet at £25 by its tenant who paid the estate only 19s 10d. When annual rents became standard later in the nineteenth century, the Landor family received about £2,000 a year in rents from 1,335 hectares of farmland and a further 1,416 hectares of mountain. As landlords, they were responsible for the maintenance of buildings (costing about £500 a year in the early twentieth century) while the tenant provided stock and equipment.

Although the ownership of land brought diminishing financial returns after the 1870s, it retained social prestige, and brought sporting rights denied to the tenants. Wealthy solicitor, Richard Baker Gabb (1840–1919), acquired his Coed-dias estate in the Grwyne Fawr from the Marquess of Abergavenny in 1885 largely for the grouse shooting and trout fishing that came with it. Other landlords, including the Bailey family, now Lords Glanusk, began to sell up. During the twentieth century income tax on agricultural rents rose

Fordson Standard tractor (perhaps the first to be used in the valley) and binder at Trefeddw, about 1920.

© *Edith Davies*

Pony and trap delivering milk in Abergavenny from Bettws Farm, 1920s.

© Jenny Parry and Smith family

dramatically and by 1953, when the rental of the Llanthony estate was £2,265, nearly all this disappeared in taxes and repairs. Not surprisingly, the estate was sold off piecemeal over the next decade.

Agricultural Depression and World War

Until the 1870s, farming incomes were increasing, and many farm buildings in the valleys were rebuilt or enlarged. Some new crops, such as flax and clover, were grown, and machinery introduced (Llwyn Celyn, Cwmyoy, had a winnowing machine by 1848) but the valleys did not see the heavy investment in 'scientific' agriculture that took place in much of England. Massive increases in grain imported from the American Midwest caused prices to crash and by 1895 cereal prices were at their lowest for 150 years. The local economy was hit even harder at the end of the century by the import of refrigerated meat and of wool from the Americas and New Zealand.

Incomes fell steeply, and many small unprofitable farms were simply abandoned. The labour force contracted, as people sought higher-paid employment in the mines and steelworks of South Wales or in the colonies. The valley's remaining farms turned increasingly to dairying, supplying milk to the growing town of Abergavenny and to the industrial valleys of west Monmouthshire. Some also bred riding horses and pit ponies.

During the First World War both men and horses were conscripted. There were attempts to increase domestic food production, repeated on a far greater scale in World War Two, when county-based War Agricultural Executive Committees directed farming by determining land usage and types of crops to be grown. Evacuees, Italian and German prisoners of war, conscientious objectors, and the Women's Land Army all worked locally in farming and forestry. After the war, agriculture was still indirectly controlled by the Government, guaranteeing a market and stable prices for agricultural products. The Agriculture Act of 1947 introduced the scheme of payments, grants, and subsidies that gradually transformed the Welsh farming economy.

Above Llanvihangel Crucorney Post Office and Forge, c.1900.

Timber and stone

Agriculture had always been supplemented by other industries.
The most important of these was forestry, and those hillsides
below about 350 metres which were too steep to farm were
managed as woodland. In 1855 the Llanthony estate offered
more than 3,000 ash trees for sale, together with oak, elm, beech
and poplar. A wood-fuelled iron forge in Llanvihangel Crucorney
was operating for much of the seventeenth century. The woods
above Cwm Coed-y-cerrig (mostly ash, beech and hazel) were
coppiced for the charcoal that gave the hamlet of Forest Coalpit
its name. This was used by the ironworks at Glangrwyney
(until its closure in 1842), as well as the smithies at Cwmyoy,
Llanthony, Llanvihangel Crucorney and Pandy, and on the
Ffwddog. Charcoal hearths extended a couple of miles up the
Grwyne Fawr, and the industry only came to an end in the late-
nineteenth century. Even then cordwood was still being taken
by mule to a chemical works in Pontrilas making naptha gas.

In 1932 the whole hamlet of Grwyne Fawr (283 hectares
of farmland, and 607 hectares of mountain grazing) was sold
to the Forestry Commission by Lord Glanusk for £6,500.
The Commission also acquired a further 142 hectares on the
Ffwddog in 1937 to create the Mynydd Du forest. In 1949 the
countryside writer H. J. Massingham described the result as
'Hades ... the original landscape has been effaced and what has

Above right Llwyncelyn,
Grwyne Fawr, in the
1960s, before its
demolition by the
Forestry Commission.

© *Isabel McGraghan*

69

taken its place [are] walls of light green and dark green in vast uniform blocks on either side of the river.'

Whatever the visual impact of its conifer plantations, the Forestry Commission did bring jobs, and built housing for its workers. In 1952 it extended its operations into the Vale of Ewyas, leasing 142 hectares above Llanthony. Today its successor, Natural Resources Wales, works through contractors, but timber production remains significant to the local economy. Stone quarrying also has a long history. The sandstone used for the buildings and boundary walls of the valleys was dug as near as possible to where it was needed, and there are many small quarry sites along the sides of the valleys, some of them worked until the twentieth century.

The Grwyne Fawr Dam

In 1912 a major civil engineering project began in the upper Grwyne Fawr. This was the building of a dam and reservoir to supply water to the mining town of Abertillery over thirty miles away. The site was at about 550 metres (1800 feet), and over twelve miles from the nearest railway station at Llanvihangel Crucorney. The work was expected to take less than four years. It required a completely new road from Stanton to Pont Escob through Cwm Coed-y-cerrig, and major work on the lanes in the Grwyne Fawr to get access to the site. Nevertheless this road was in places too steep even for traction engines, and a railway had to be built along it.

Around 400 men worked on the dam and pipeline, and an entire village of corrugated iron bungalows, known as 'Navvyville' or 'Tin Town', with its own school, canteen, mission hall and hospital, was built at Blaen-y-cwm. Work stopped at the end of 1915, and was resumed after the First World War. Five small steam engines operated on the single-line track in 1923, and two more were acquired in 1925. The stone and concrete dam, two miles above 'Tin Town', is one of the tallest in Britain, and work was only completed in March 1928. 'Tin Town' and the railway were then dismantled, but the new road remained, opening up the valley for the afforestation that followed.

Above The corrugated iron village at Blaen-y-cwm, about 1920.

Left The Grwyne Fawr dam, completed in 1928.

Recovery and Diversification

After World War Two, the Hill Farming Act of 1946 offered grants for rehabilitation of sheep holdings, though much stock was lost in the terrible winter of 1946–7. The 1947 Agriculture Act introduced hill sheep and cattle subsidies, which were extended in later legislation. Some buildings were repaired (often with concrete blocks and asbestos sheet), but most were unsuitable for mechanised farming and larger steel-frame sheds were built. More stock was kept in the 1950s – typically about

500 sheep and thirty cattle on a fifty-hectare farm. Tractor-ownership, rare before 1939, became almost universal in the mid '50s, less and less corn was grown, and by the 1960s arable farming, other than root crops, had almost disappeared.

When Britain joined the European Economic Community in 1973, rural Wales initially benefitted from 'less favoured area' status. During the 1980s conditions became more difficult, with cuts in support and quotas on production, a development compounded in the '90s with the regulations and export restrictions resulting from 'mad cow' disease in cattle. Nearly half of all farms were then described as 'unsuccessful' and it was no longer possible to generate an adequate income from a fifty-hectare holding. Consequently many farming families became dependent on a second external income.

Today, the role of the agricultural sector remains important, but tourism has grown greatly since the 1960s. There is increasing debate about how the land should be managed. It seems likely that farming will become less intensive, as land is managed for its ecological and amenity value.

A Ferguson TE20 tractor at Llanthony, about 1950.

© *Christine Olsen*

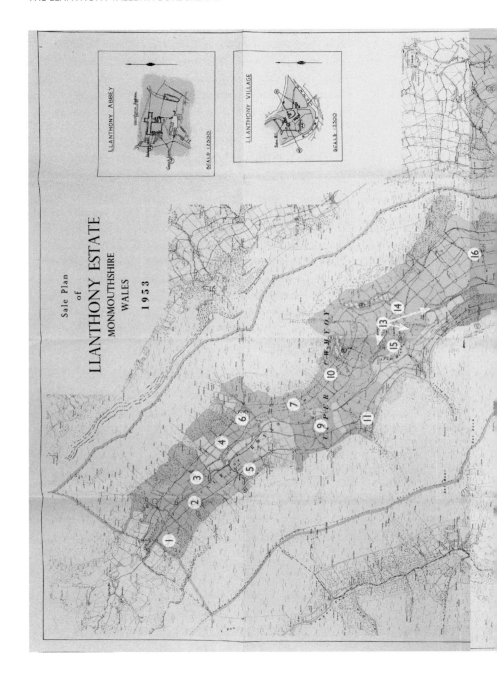

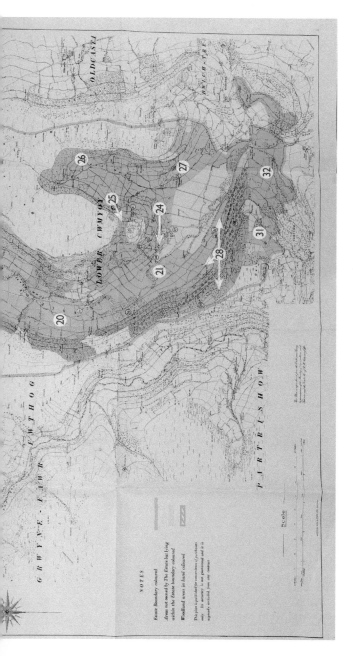

The Farms of the
Llanthony estate in 1953.

1 Vishon	6 Trevelog	13 Abbey Holding
2 Pen-y-wyrlod	7 Llwyn-on	14 Court Farm
3 Garnfawr	9 Nant-y-carnau	15 The Mill
4 Ty-hwnt	10 Broadley	16 Maes-y-Beran
5 Troedrhiwglas	11 Nant-y-Gwyddel	18 Lower Henllan

19 Upper Henllan	26 Blaenyoy
20 Neuadd Llwyd	27 Perth-y-crwn
21 Neuadd	28 Pen-yr- heol
24 Pen-y-wern	31 Gaer Farm
25 Ty-hwnt-y-bwlch	32 Llwyn Celyn

Areas in yellow in other ownership. Numbered sites edited for relevance.

Gwent Archives

75

4

Buildings in the Landscape

The Black Mountains have a rich heritage of vernacular buildings, some dating back to at least the sixteenth century and still occupied. These were originally the houses and farmsteads of the area's more prosperous inhabitants. By contrast the cottages of the poor built before the nineteenth century rarely survive as more than a ruin.

Hall houses

In the fifteenth century, as the area recovered from the devastation of the Glyn Dŵr rising, new houses were built to replace those destroyed during the conflict. Most were timber-framed (built mainly of wood and earth) with a thatched roof supported by a pair of great curved oak trusses known as crucks. Their main living space was the hall, a large room open to the roof and with a hearth in the middle of the floor. None of these timber-frame houses survive unaltered, though a few still retain crucks and smoke-blackened rafters, hidden above later ceilings.

People began to build in stone during the fifteenth century, as the local sandstone was easy to quarry, and lime for mortar was also available. A stone-built house, with high-quality internal carpentry, seems to have been a mark of wealth and local importance. The earliest example in the area is Llwyn Celyn, Cwmyoy, built around 1420. Others include the Neuadd at Partrishow and the hall range of Little Llwgwy in Cwmyoy.

Early-eighteenth century combination barn and byre at Ty mawr, Ffwddog, built down the slope with the cattle at the lower level.

© Simon Powell

LLWYN CELYN, CWMYOY

Llwyn Celyn (meaning Holly Bush or Grove), built in 1420 on priory land, is one of the oldest surviving domestic houses in Wales. Standing on a steep south facing slope, it commands the southern mouth of the Llanthony Valley. As originally built, it was an open hall house with a cross passage and service rooms, and a two-storey solar range at the high end of the hall to provide a parlour and chambers above. An arched door from the hall probably led to an external stone staircase (now lost) up to the first floor chamber. That the house was of high status is confirmed by Llwyn Celyn's exceptionally fine joinery: it has ogee-headed doors to the service rooms and parlour, the remnants of a spere truss (a decorative roof truss at the low end of a hall), and a fixed bench at the high end of the hall, where there is also evidence of a dais canopy. It is not yet known why or by whom Llwyn Celyn was built, but it seems sure that its first inhabitant was a high ranking official associated with the priory, and perhaps even the prior himself.

Above Llwyn Celyn has a commanding position at the southern end of the valley. Clustered around the early fifteenth-century house are a later threshing barn and other farm buildings, most now converted to community use.

Ground floor plan.

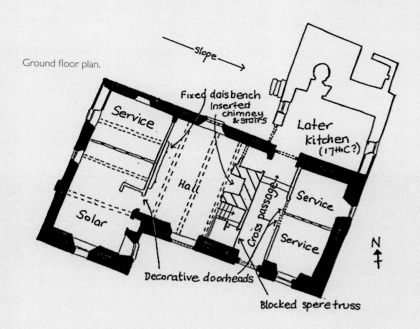

Service

Fixed dais bench
Inserted
chimney
& stairs

Later
kitchen
(17th C?)

Slope →

Hall

Cross passage

Service

Service

Solar

N

Decorative doorheads

Blocked spere truss

Below Llwyn Celyn is rare in having a cross wing (left) built at the same time as the main hall, its upper floor also originally open to the roof. The dormer is a later insertion added to light the upper floor when the hall was ceiled over, probably around 1650.

The house has been known as Llwyn Celyn since at least 1597, and it was the home of well-to-do copyhold tenants of the Llanthony estate, members of the Watkins and George families, until 1762. Such tenants added a rear kitchen, and in the seventeenth century, inserted a ceiling across the space of the hall to create an upper floor, accessed by a wooden dogleg staircase. At the same time, a massive chimney stack was inserted to replace the open fire. The site also acquired the outbuildings typical of a self-sufficient valley farmstead: beast house or byre, threshing barn, cider house, malt and wheat kilns and pigsty.

Below The ground floor of the hall, which was once open to the roof timbers. The blocked, pointed doorway was part of the 1420 house and led to a staircase leading to the upstairs of the cross range. This solar range is also entered through the adjacent door which has another finely carved ogee doorhead. The fixed bench along the wall may also date to 1420, although the panelling is later.

Above Llwyn Celyn was completely re-roofed in diminishing courses, using stone tiles quarried at Longtown in the adjacent Olchon Valley. This provided an opportunity for training in traditional Welsh roofing techniques.

Left One of the pair of finely carved ogee doorheads in the cross passage. These doors led to service rooms but would have been in full view of the high table through the opening in the spere truss, later filled by the inserted chimney stack.

The house itself remained unchanged through these centuries of tenant farming and eventual decline. In the 1850s, the Jasper family took on the lease, and it was Jasper relations, Tom Powell and his wife Olive, who bought the farm from the Llanthony Estate in 1959.

Such is the importance of the site that in 2012, by now in advanced decay, Llwyn Celyn was acquired from their sons, Trefor and Lyndon Powell, by national building preservation charity, the Landmark Trust, with funding from Cadw and the National Heritage Memorial Fund. Comprehensive restoration followed from 2016–18, with enabling funding from the Heritage Lottery Fund. Today, Llwyn Celyn is let to all for holidays by Landmark, with a community centre in the threshing barn and an information room in the beast house.

Right The inserted floor that made the space of the open hall into two floors. Originally, the ground floor hall was open to these fine roof timbers, now revealed again.

Most of these hall houses now have a later chimney and upper floor over the hall. They usually had a room at one end of the hall which was the owner's private space, and some sort of service area at the other, separated from the hall by an entry passage running across the house.

Longhouses

Farm incomes were rising in Britain in the sixteenth century, and much of this prosperity was spent on building new houses and modernising older ones – often with a complete upper floor and with several heated rooms. This wave of rebuilding seems to have started relatively late in the Black Mountains – towards the end of the sixteenth century – and to have continued into the 1640s.

Neuadd, Partrishow, a stone hall house, probably built around 1500. The dormer windows and chimney are later, and the door into the cross-passage is on the right.

Below Bridge Cottage, Llanvihangel Crucorney, probably built in the early seventeenth century, with the door into the hall at the gable end. Photographed in the 1940s before it was enlarged.

© Amgueddfa Cymru – National Museum of Wales

Houses of this period have a standard floorplan, found in much of Monmouthshire and Breconshire. Their halls have a fireplace backing on to the cross passage, and two little rooms beyond a panelled oak partition. There are one or two bedrooms on the floor above. These sometimes had attics over them. Some houses were entered through a door in the gable end – for example Bridge Cottage (Llanvihangel Crucorney) – but others had a cross passage. Much of people's wealth was tied up in their cattle, and as cattle thieving was endemic in the Marches the more valuable beasts were stalled in a byre attached to the house.

These 'longhouses' were usually built down a slope with the byre at the lower end. Examples include Dan-y-bwlch, as well as Coed Farm (Partrishow), a timber-framed house later encased in stone. Unusually the byre at the

Dan-y-bwlch, between
Cwmyoy and Bwlch
Trewyn, a sixteenth-
century longhouse,
photographed during
recent restoration. The
byre is on the right.

© *Crown copyright: Royal
Commission on the Ancient and
Historical Monuments of Wales*

Ty-hwnt-y-bwlch, Cwmyoy, another sixteenth-century longhouse, showing
the byre (left) and the seventeenth-century parlour extension (right).

© *Simon Powell*

Little Llwgwy, Cwmyoy,
with a parlour tower
(left) of about 1610,
added to an early
sixteenth-century hall
house.

© *Royal Commission on
the Ancient and Historical
Monuments of Wales*

Coed still housed cows until the 1970s, as from the late
1600s cattle were usually stalled elsewhere and the byres of
most longhouses were rebuilt to provide additional living
space. Ty'n-y-llwyn (Partrishow), built in 1598–9, was
remodelled in this way in 1649.

Ty mawr, Ffwddog, the
surviving parlour wing,
built around 1630, of
an older hall house.

© *Simon Powell*

Parlours and Privacy

From the early-seventeenth century people placed greater
value on privacy, and had more money to spend on
household goods. As well as their hall, where the household
ate, they needed another large room
for entertaining friends, usually called
a parlour. Sometimes there was space
enough to create a parlour in a new
wing at the upper end of the hall,
but the byre at the lower end was
often rebuilt as a parlour, with a best
bedchamber on the floor above. These
houses emulate in miniature two gentry
mansions nearby, Old Gwernyfed
(Velindre), and Llanvihangel Court
(Llanvihangel Crucorney).

TY'N-Y-LLWYN

Ty'n-y-llwyn (the house in the grove) is built on a steep slope, with a retaining wall at the front. Its two-and-a-half-storey hall range was built in 1598–9 (felling dates of the timbers obtained through tree ring analysis). On the ground floor there is a small parlour and service room beyond the hall. A stone spiral stair leads to a first-floor chamber with a fireplace and privy, and there is a large attic above. The house was originally entered through wide doors at either end of a cross passage. On the downhill side of the cross passage was a byre with a loft above it.

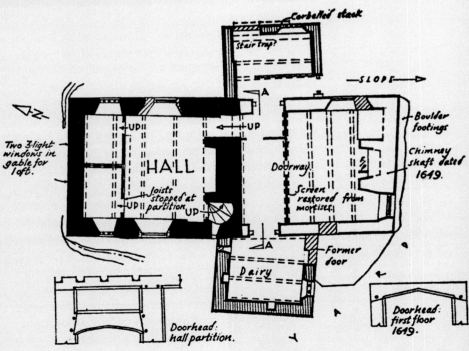

Ground-Floor Plan of Ty'n-y-llwyn (from *Brycheiniog*, XII, p. 33).

Above The late-Elizabethan hall range from the west showing the steep downhill slope and the later cross wings.

© *Simon Powell*

Above right The hall, looking towards the screen which divides it from a small parlour and store room. The windows were originally unglazed and had internal shutters to keep out the worst of the weather.

Right The byre range (left) was rebuilt in 1649 to provide a parlour. The prominent diamond-shaped chimney stacks were one of the hallmarks of a gentry house.

© *Simon Powell*

In 1649 (date on the chimney stack) the byre end was rebuilt to provide a heated parlour and another large bedchamber on the first floor. Two little wings were also added on either side of the house, making it cross-shaped in plan. A number of cross-shaped houses were built in the Marches in the mid-seventeenth century, and the smaller arms of the cross usually provided a porch and a staircase. Here, however, the eastern wing seems to have accommodated a second semi-independent household. One of the largest freehold farms in the Grwyne Fawr, Ty'n-y-llwyn was the home of the Powell family from the eighteenth century to 2003.

THE LLANTHONY VALLEY: A BORDERLAND

Some of the seventeenth-century parlour extensions to older houses were taller and more impressive than the hall range. The parlour range at Ty mawr (Ffwddog), probably built in the 1630s, survived the later demolition of the house's sixteenth-century open hall. Little Llwgwy also has an added parlour tower of about 1610. Penyclawdd Court (Llanvihangel Crucorney), probably built in the early sixteenth century, was modernised a hundred years later when a tall parlour wing with a stair tower was added.

These new parlours had larger, glazed windows, and were more expensively furnished. In many houses, the upper floor was reached via a stone spiral staircase built into the wall beside the hall fireplace. Later the stone treads are replaced by broader oak ones, and by 1700 framed wooden staircases of the sort inserted into the earlier hall at Llwyn Celyn had become usual. Houses often accommodated more than one generation – widows sometimes had the right to a chamber of their own – which meant more private rooms were needed.

Cooking on an open fire at Ty-Shores, Llanthony, in about 1950.

© Christine Olsen

Trewyn, late medieval in origin, rebuilt in 1692 with a symmetrical front and hipped roof.

© Simon Powell

Where did they cook?

Every household needed a fire for boiling and roasting, and an oven for baking. Most people cooked and ate in their hall, but some seventeenth-century inventories list a separate kitchen in the larger houses. Kitchens might be in the house itself, or in a detached building a few yards away, reducing the risk of fire. Some of these external kitchens survive – at Cwm-bwchel (Llanthony), Upper House (Ffwddog), and Coed (Cwmyoy).

In the eighteenth century, unless a new kitchen was provided, the hall or the parlour often became a large kitchen-living room. People cooked over wood or peat fires until coal-fired ranges were installed in the second half of the nineteenth century. Baking often continued to be done in an outbuilding. A wood fire was lit in the oven, and when it was hot enough, the ashes were raked out, and the loaves baked on the hot stone.

Change and Continuity

By the late seventeenth century any new gentry house, even in the Welsh borders, was a square or rectangular building with a central doorway and large symmetrically arranged windows

The barn at Trefeddw, built around 1700, with opposed
doors to the threshing floor and a small byre at one end.

© Simon Powell

The byre at Ty'n-y-llwyn, probably built in the late-seventeenth century. The central door to the feeding passage is now a window.

© *Simon Powell*

Below The mill (right) and miller's cottage at Llanvihangel Crucorney in about 1900. Powered by an overshot waterwheel, this continued to grind corn for local use until the 1930s.

© *John Evans*

– examples include Trewyn (Bwlch Trewyn), rebuilt by the Delahaye family in 1692. For the first time, the front of these houses became more important than the back. They were built across the slope, with the more important rooms looking outwards into the landscape.

Below this level of society, the old ways persisted. Pont-Rhys-Powell Farm (Cwmyoy) was remodelled with a cross passage as late as 1689. During the eighteenth century some houses were re-roofed, or given more symmetrical facades (as at Court Farm, Llanthony), and re-planned internally, but attention now turned to the outbuildings that surrounded every farm.

Barns, Byres and Mills

The valley's mixed farming economy required a range of specialist buildings. Originally timber-framed, these outhouses were rebuilt in stone after about 1650. Every farm had a barn to store the harvest, with a paved threshing floor between wide doorways which provided a through draught. Some, such as the late-seventeenth century barn at Trefeddw, dwarf the houses they serve.

Cattle were stalled in the byre, which often had three entrances at one end, the central one leading to a feeding passage, as the cows were tethered facing each other across the building. From the end of the seventeenth century, barn and byre were often combined in a single large building constructed down a slope, with the cattle stalled at the lower level, as Ty mawr (Ffwddog). There are dated examples at Pont-Rhys-Powell (1703) and Ty-hwnt-y-bwlch (1720). On the larger farms a malt house was used for malting barley, and for drying grain before it was stored in the granary.

Horses were stalled in the stable, but if a stallion or a bull was kept, these had to be housed separately. Other smaller structures around the farmhouse included cart sheds, calves' cots, pigsties, goose cots, even bee boles. Water mills were another feature of the landscape. None of those on the Honddu and Grwyne Fawr survive in recognisable form, but the out-buildings of Olchon House Farm include a small working water mill dating from the mid-nineteenth century.

Above The mid-nineteenth century overshot waterwheel at Olchon House Farm, Llanveynoe.

© Brian Dixon

STUDYING LOCAL HOUSES

The first systematic study of Monmouthshire house was carried out by Sir Cyril Fox, director of the National Museum of Wales, and Lord Raglan, between 1941 and 1948. A number of houses in Cwmyoy and Llanvihangel Crucorney were included in their three-volume *Monmouthshire Houses*. Meanwhile vernacular houses in Breconshire were surveyed in the 1960s by Stanley Jones and John Smith, who published those in the Crickhowell District in *Brycheiniog* in 1967. Some of these houses are also discussed in the *Monmouthshire* and *Powys* volumes of *The Buildings of Wales* series inspired by Nikolaus Pevsner.

Today descriptions and photographs of the area's listed buildings can be found at www.coflein.gov.uk. However some pre-1700 houses remain unlisted, and much work is still needed on farm buildings.

Lord Raglan, photographed by Sir Cyril Fox at Lower Trefeddw, c.1945.

© *Amgueddfa Cymru – National Museum of Wales*

Nantybedd, Grwyne Fawr, built by the Forestry Commission in 1934.

© *Simon Powell*

Nineteenth-century repair and improvement

Many of the houses in the Llanthony Valley were enlarged or rebuilt by the Landor family during the nineteenth century. Typically the roof was raised, providing more space in the attics, and covered with Welsh slate (replacing heavy sandstone tiles). Larger windows were made, often with metal frames. Old houses modified in this way include Court Farm and Trevelog (Llanthony). Cottages in Llanthony village with central porches and picturesque dormer windows were probably built around 1840.

In the second half of the nineteenth century, although the Llanthony and Glanusk estates repaired their properties, few new houses were needed. Nevertheless, until farm profits collapsed at the end of the century, improvements continued to be made to farm buildings, sometimes using corrugated iron, and additional field barns were built.

A cottage in Llanthony, probably built by the Landor estate around 1840.

© *Christine Olsen*

Public Housing

After purchasing much of the upper Grwyne Fawr in 1932, the Forestry Commission built a number of new houses for its workers. These include Nantybedd. In 1952 the Commission built a pair of new houses on the edge of Llanthony village. These were good examples of well-planned post-war public housing, and Abergavenny Rural District Council also built council houses in Crucorney and at Stanton (1959). Some older houses in the valley were demolished in the '50s, but from 1952 others began to be listed under the Town and Country Planning Act of 1947 as being of especial historic interest. The Brecon Beacons National Park, with its mission to protect the landscape, became the planning authority ten years later.

Since the 1980s, many older properties have been restored, and in some cases extended, while some agricultural buildings have been converted for domestic use. As farms have expanded in size, houses have often been separated from much of their land. Although there is still more to discover, the built heritage of the valleys is probably now better understood and valued than at any date in the past.

5

A Place of Inspiration

The Llanthony Valley has a unique sense of place. In the late eleventh century William de Lacy was so moved by St David's ruined chapel that he became a hermit and inspired the founding of Llanthony Priory. Over the last two hundred and fifty years in particular, since we began to find beauty in the wild, artists and writers have found inspiration in the valley – more perhaps than anywhere else in Wales.

Romantic Tourists

Before the early eighteenth century, hardly anyone travelled for pleasure. The few who wrote about their experiences were horrified by upland Wales – by its bad roads, miserable inns, poor food, incomprehensible language and terrible weather. But by the 1750s, attitudes were changing. People began to find pleasing qualities in nature, and to admire landscapes they felt to be 'picturesque' and that reminded them of paintings by Claude Lorrain or Salvator Rosa. Antiquarians also started to rediscover the history and culture of Wales, and were fascinated by the ruined castles and abbeys that recalled its medieval past.

The brothers Samuel and Nathaniel Buck included an engraving of Llanthony Priory in their *Venerable Remains in England and Wales*, published in 1732. From the mid-eighteenth century travellers could take a regular boat service down the river Wye, from Ross to Chepstow.

Opposite Eric Gill at Capel-y-ffin in about 1928, photographed by Howard Coster.

© *National Portrait Gallery, London*

Following pages The Ruins of Llanthony Priory in 1732, drawn and engraved by Samuel and Nathaniel Buck.

© *Amgueddfa Cymru – National Museum Wales*

THE NORTH WEST VIEW OF LANTON

To Edward Harley Esqr.
Knight *of the Shire for the County of Hereford*
Proprietor of these Remains
This Prospect *is gratefully inscrib'd by*
his most oblig d humble Servts. Saml. & Nathl. Buck.

BY, IN THE COUNTY OF MONMOUTH.

THIS PRIORY or as y. Welsh call it Llandevi Nanthodeny, from an Ancient little Chapple which stood here on y. River Hodeny. The solitariness of this Place made S.t David build here his little Hermitage, in which we do not find any succeeded him, till about y. Year 1103. William, a Knight, of y. Family of Hugh de Lacy retir'd to an austere Life alone in it, but at length Er'nesius Chaplain to 2. Maud became his Companion, & were both of them remarkable for their simplicity of Manners & Sanctity, which made Hugh de Lacy offer them many large Gifts, all which they refus'd except enough to build them a small ordinary Church w.ch was dedicated to S. John Baptist An. 1108. In Process of Time Er'nesius by y. Advice of Anselm A.ch. B.p of Canterbury perswaded W.m to accept of some of these generous offers & form a Convent, which was done, & they chose y. order of Canons regular of S. Austin.——— An. Val. Dug. £9 19 0.
S. & N. Buck del. et fc. 1732.

94212

Travellers marvelled at the ruins of Goodrich Castle and Tintern Abbey. Britain's long war with France between 1792 and 1815 made European travel impossible, and so a tour round Wales became increasingly popular with the leisured classes.

Tourists started coming to the Black Mountains during the 1770s. Some wrote about what they saw, and employed artists to record it. They include Henry Wyndham (1736–1819) who travelled in Wales in 1774–75, and in 1777 when he was accompanied by the Swiss watercolourist Samuel Hieronymous Grimm. Wyndham wrote that the Priory 'offers as picturesque and romantic remains as are to be seen in any part of the tour'. Grimm's watercolours are set in a sunlit landscape, 'enriched with meadows and corn fields, and now and then enlivened with a little wood'.

Other artists to paint Llanthony in the 1770s and 1780s include Paul Sandby, William Hodges and John 'Warwick' Smith.

Opposite Llanthony Priory painted by William Hodges in 1777. The crossing tower was reduced in height in 1808 and the south aisle (left) collapsed in 1837.

© Amgueddfa Cymru – National Museum Wales

Below Tourists and locals: the west front of the Priory painted in watercolour by Samuel Hieronymous Grimm in 1777.

© Amgueddfa Cymru – National Museum Wales

A later more dramatic view by Turner of around 1834, engraved in 1836 for the *Picturesque Views in England and Wales*.

© *Indianapolis Museum of Art, Bequest of Kurt F. Pantzer*

Opposite A distant
prospect of the Priory
by the seventeen-year-
old J.M.W. Turner.

© *Indianapolis Museum of
Art, Gift in memory of Dr. and
Mrs. Hugo O. Pantzer by their
Children*

J.M.W. Turner first came in 1792, when he was only
seventeen. The antiquarian and amateur artist Sir Richard Colt
Hoare (1758–1838) repeatedly sketched the Priory, and made
illustrations for *An Historical Tour in Monmouthshire*, published
in 1801 by William Coxe (1748–1828). Coxe describes
venturing up the valley:

> 'Except in very few places, there is not room for a single
> horse to pass a chaise; and should two carriages meet, neither
> could proceed until one was drawn backwards to a considerable
> distance. The soil is boggy in wet, and rough in dry weather;
> the ruts made by the small Welsh carts are extremely deep, and
> sometimes we were prevented from being overturned only by the
> narrowness of the road and the steepness of the sides...
>
> 'The vale itself is fertile in corn and pasture [but]... is wholly
> encircled by an amphitheatre of bleak and lofty mountains ...'

William and Dorothy Wordsworth came to Llanthony in
1798, and increasingly visitors to the Black Mountains thrilled
at the bleakness of the hills – 'the rocky peaks of the Black
Mountains, over which the foot of man has scarcely ever
trod' as Thomas Roscoe wrote in 1837. A later watercolour of
Llanthony made by Turner around 1834 emphasises the height
of the hills, and the dramatic rain-filled clouds.

The Priory remained a popular destination (from 1864
visitors could take the train as far as Llanvihangel Crucorney)
even though the collapse of four piers of the south arcade in
1837 narrowly missed a picnic party. In 1870 the diarist and
clergyman Francis Kilvert found 'two tourists ... postured
among the ruins in an attitude of admiration, one of them of
course discoursing learnedly to his gaping companion and
pointing out objects of interest with his stick'. Today in the
summer he might find two hundred.

Eric Gill and David Jones

In 1924 the sculptor, engraver and letter cutter Eric Gill (1882–1940) left the Catholic art-and-craft community at Ditchling, Sussex, and moved with his family and a few friends to Father Ignatius's former monastery at Capel-y-ffin. There he was to make some of his finest sculptures, among them *The Sleeping Christ*, while carving the tombstones and inscriptions that kept him solvent. During the 1920s Gill developed new typefaces for the Monotype Corporation intended for machine production. These types, Perpetua (1925), Gill Sans (from 1927), and Solus (1929) are still in use today. This text is set in Gill's Joanna typeface (1931).

Gill loved Capel-y-ffin – where he could bathe naked in the mountain pools, climb the hills and go on picnics. However he was often away on business, and his rural idyll was dependent on the work of his wife and daughters. Like their neighbours, they had to run a farm, and do their own baking, brewing, milking and butter-making. There were almost no cars in the Black Mountains in 1924, so the community used pony carts to make the fifteen-mile journey to Abergavenny. Gill later bought

Sleeping Christ, carved by Eric Gill in Caen stone in 1925.

© *Manchester City Art Galleries*

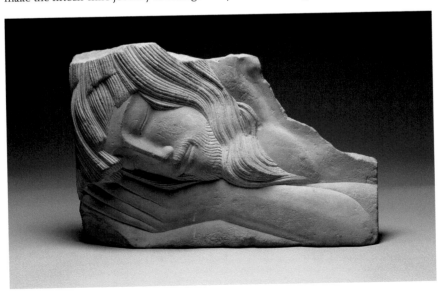

the monastery, but its isolation became too much and in October 1928 he left Capel-y-ffin for Pigotts, in the Chilterns.

At Ditchling and at Capel-y-ffin Gill gathered disciples around him, often men who had been through the First World War. They included the London-Welsh artist and poet David Jones (1895–1974). Jones was for a time engaged to Gill's daughter Petra, and at Capel-y-ffin he was to find his direction as an artist in the 'strong hill-rhythms' of the valley. Works like his *Capel-y-ffin* (1926–7) are significant landmarks in British modern art, and have a crispness of form and unsettling construction of space that recall the landscapes of Cézanne. Elements drawn from the valley, the rushing streams, the drystone walls, and the hill ponies – to him the descendants of the steeds ridden by Arthur's knights – reoccur in his work years later.

David Jones's Christmas present to the Gill family in 1926, this watercolour takes the viewer up the hillside from the Monastery towards the Grange.

Amgueddfa Cymru –
National Museum Wales
© the Estate of David Jones

Neo-romantic Artists

Eric Gill's years at Capel-y-ffin inspired other artists to paint in the valley during the 1930s and 1940s. These included Cedric Morris (1889–1982) who painted at Llanthony in 1935. Eric Ravilious (1903–1942) had worked mainly in Sussex, but early in 1938 he came to the valley in search of a wilder landscape. He spent several weeks at Capel-y-ffin, and found 'the hills … so massive that it is difficult to get them down on paper'. His watercolours of the valley in winter include The Waterwheel.

John Piper (1903–1992) visited Ravilious, and in 1941 returned to paint the Priory. Piper's work during the '40s was an emotional response to the British landscape and history, reflecting the trauma of the Second World War.

During his stay at Capel-y-ffin early in 1938, Eric Ravilious painted The Waterwheel, a home-made contraption of wood and petrol-tins in the Honddu powering a grindstone.

© Brecknock Museum and Art Gallery

In 1941, John Piper stayed at the Abbey Hotel, and this oil
painting depicts the ruins in blocks of contrasting colour.

Amgueddfa Cymru – National Museum Wales © DACS Estate of John Piper

John Craxton's ink and watercolour *Llanthony Abbey*,
1942, contains dramatic contrasts of light and shade.
This combination of emotion and nature was central to
the British Neo-Romantic movement of the 1940s.

Tate © Estate of John Craxton

Right Wood engraving *The Monastery, Capel-y-ffin*, made by Edgar Holloway in 1949.

Gwent Archives © Estate of Edgar Holloway

Another 'neo-romantic' was John Craxton (1922–2009) who came to Llanthony in 1942. He found 'crows and jackdaws nesting in the broken gothic windows, ivy everywhere'. A sombre drawing frames the ruins between the roots of a fallen tree.

In 1943 the watercolourist and printmaker Edgar Holloway (1913–2008) stayed at the monastery at Capel-y-ffin, now run as a guesthouse by Eric Gill's daughter Betty. There he met and married Daisy Monica Hawkins, the model for many of Gill's life drawings. They rented the Sychtre Farm for a while from the illustrator Reg Gammon (1894–1997) and lived intermittently at Capel-y-ffin until 1950. Gammon had moved to Llanthony at the start of World War II, and became a hill farmer. He stayed for twenty years, making images of life in the valley, the first artist to become a participant, rather than a passing observer.

James Wood (1889–1975), painter, writer and aesthete, owned a cottage in Cwm Siarpal, above Llanthony. He and his sister Lucy Boston (author of the Green Knowe children's books) were friends of the textile designer and painter Elizabeth Vellacott (1905–2002). She too came to Llanthony, making an intense study of the landscape. Her compositions are formed from thousands of tiny pencil or chalk marks. Edward Burra (1905–1976), better known for his surreal images of the urban underworld, was to paint at Llanthony in 1968.

Mountain Top – The Gully, Llanthony, a drawing made by Elizabeth Vellacott in 1966.

Kettle's Yard, University of Cambridge
© Estate of Elizabeth Vellacott

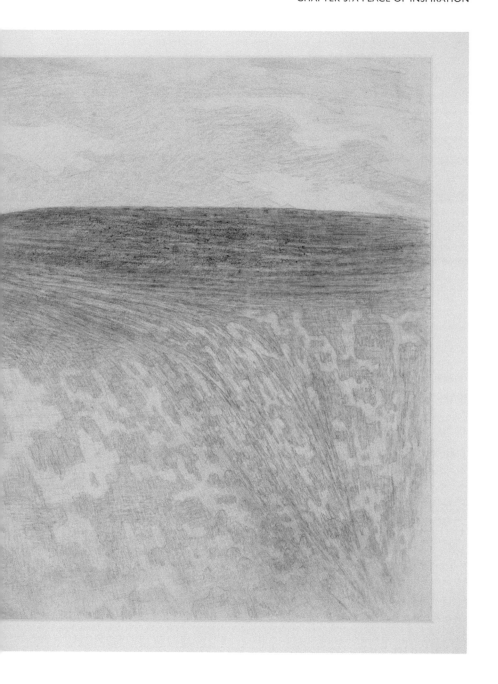

From Ley-lines to LSD

A new generation of writers explored the Welsh Borders after the First World War. Llanthony was a key site for Alfred Watkins (1885–1935) who constructed his own theory of pre-history based on the network of straight trackways he found in the landscape, which he interpreted as channels of energy and called ley-lines. Raymond Williams (1921–1988) was brought up in Pandy, where his father worked on the railway. Remembered today for *Culture and Society* (1958) which explores the meaning that British writers had given to the word 'culture' since industrialization, Williams was also a novelist and became professor at both Cambridge and Stanford universities. He wrote of the world of his childhood in *Border Country* (1960). The travel writer Bruce Chatwin (1940–1989) came often to Capel-y-ffin as a boy and it helped shape his 1982 novel *On The Black Hill*. He fictionalised the hill farm in the story, but borrowed its name from the Vision near Capel-y-ffin. Much of the book was written while staying with his publisher Tom Maschler near Llanthony. Maschler brought an even more unlikely visitor, the beat poet Allen Ginsberg (1926–1997) in 1967. While on LSD Ginsberg began 'Wales Visitation', a poem on the sublime in the Romantic tradition of Wordsworth which ends:

> *Heaven breath and my own symmetric*
> *Airs wavering thru antlered green fern*
> *drawn in my navel, same breath as breathes thru Capel-Y-Ffin,*
> *Sounds of Aleph and Aum*
> *through forests of gristle,*
> *my skull and Lord Hereford's Knob equal,*
> *All Albion one.*

The continuing tradition of the Romantic landscape still brings painters, photographers and film makers to the Black Mountains today. Robert Plant of Led Zeppelin filmed here. Iain Sinclair's work of literary archaeology *Landor's Tower* (2001) draws on local spirits, and Owen Sheers's *Resistance* (2007) a counter-factual allegory of war-time German invasion, is set in the Olchon valley.

Allen Ginsberg near Llanthony in 1967.

© Tom Maschler

Index

Find out more

1. Unpublished

Records and research notes at the National Library of Wales, the Gwent Archives and the Crickhowell Archives Centre.

2. Published

Abergavenny Museum, *Sites of Inspiration: Artists of the Llanthony Valley*, exhibition catalogue, 2014

Hugh Allen, *New Llanthony Abbey: Father Ignatius's Monastery at Capel-y-ffin*, Bishop's Castle: YouCaxton, 2016

Joseph Bradney, *A History of Monmouthshire Volume 1 Part 2a The Hundred of Abergavenny* (Part 1), London: Academy Books, 1991 (1st edition 1906)

Richard Baker Gabb, *Hills and Vales of the Black Mountain District*, Hereford: Lapridge, 1993 (1st edition 1913)

Madeline Gray and Prys Morgan (eds), *The Gwent County History. Volume 3. The Making of Monmouthshire, 1536–1780*, Cardiff: University of Wales Press, 2008

Ralph A. Griffiths and others (eds), *The Gwent County History. Volume 2: The Age of the Marcher Lords, c.1070–1536*, Cardiff: University of Wales Press, 2008

Christopher Hodges, *Derelict Stone Buildings of the Black Mountains Massif*, Oxford: Archaeopress, 2015

Gordon Hopkins, *Llanthony Abbey and Walter Savage Landor*, Cowbridge: D. Brown, 1979

Jonathan Mullard, *Brecon Beacons*, London: William Collins, 2014

John Newman, *The Buildings of Wales: Gwent/Monmouthshire*, Harmondsworth: Penguin Books, 2000

Frank Olding, *Discovering Abergavenny: Archaeology & History*, Abergavenny Local History Society, 2012

Robert Scourfield and Richard Haslam, *The Buildings of Wales: Powys*, New Haven and London: Yale University Press, 2013

Priscilla Flowers Smith, *The Hidden History of Ewyas Lacy*, Almeley: Logaston Press, 2013

David Tipper, *Stone & Steam in the Black Mountains*, Abergavenny: Blorenge Books, 1985

Articles in *Beacon, Brycheiniog, Gwent Local History, Monmouth Antiquary, Valley Views*

3. Websites

British Listed Buildings (www.britishlistedbuildings.co.uk)

Coflein: The online catalogue of archaeology, buildings, industrial and maritime heritage in Wales (www.coflein.gov.uk)

Ewyas Lacy Study Group (www.ewyaslacy.org.uk)

Historic Wales (www.historicwales.gov.uk)

The Landmark Trust (www.landmarktrust.org.uk)

Llanthony Valley & District History Group (www.llanthonyhistory.wales)

People's Collection Wales (www.peoplescollection.wales)

Places of Wales [tithe maps] (places.library.wales)

Welsh Newspapers Online (newspapers.library.wales)